Contents

Foreword
Glossary

Foreword

Assessment of Water Quality by Principal Component Analysis; this book is aimed at raising awareness of researchers, scientists and engineers on the benefits of Principal Component Analysis (PCA) in water quality data analysis. In this book, the reader will find the applications of PCA in water quality research fields

This book provides a unique insight into the problems our planet faces in terms of water quality and quantity, and what to do about it. This is the only books expressed comprehensive and interdisciplinary focus to statistical understanding with the multidimensional approach.

This book made of 06 years consistently research on water resources, makes it ideal source for students, teachers, industrialist, water experts and environmentalists.

This book provides an essential guide to researchers, it offers: various aspects of PCA; on the challenges and experiences in present scenario.

Simply explained, this is an important book for all who wish to make a difference in how to plan and manage our water resources.

Dr. Hemant Pathak

M.Sc. (Gold medalist), Ph. D.

Assistant Professor of Engineering Chemistry

Indira Gandhi Govt. Engineering College

Sagar, MP, India

Glossary

- **Eigen values** measure the amount of the variation explained by each PC and will be largest for the first PC and smaller for the subsequent PCs.

- **Eigen value greater than 1** indicates that PCs account for more variance than accounted by one of the original variables in standardized data. This is commonly used as a cutoff point for which PCs are retained.

- **Eigenvectors** provides the weights to compute the uncorrelated PC, which are the linear combination of the centered standardized or centered un-standardized original variables.

- **Scree Test** Plotting the eigen values against the corresponding PC produces a scree plot that illustrates the rate of change in the magnitude of the eigen values for the PC.

- **First Principle Component** The first Principle Component is the eigenvector of the covariance matrix with the largest eigen value. This eigenvector points in the direction of maximum dispersion of the data set.

- **Second Principle Component** it is simply the eigenvector that is orthogonal to the first Principle Component.

- **PC loadings** are correlation coefficients between the PC scores and the original variables.

- **PC scores** are the derived composite scores computed for each observation based on the eigenvectors for each PC.

- **Means of PC scores are equal to zero**, as these are the linear combination of the centered variables.

- **Bi-plot** display is a visualization technique for investigating the inter-relationships between the observations and variables in multivariate data. To display a bi-plot, the data should be considered as a matrix, in which the column represents the variable space while the row represents the observational space.

Assessment of Water Quality by Principal Component Analysis

1. Introduction

Environmental analysis is innately multi-dimensional. When measuring environmental variables such as water quality uses multiple measurements. Ground water and surface water systems such as lakes, rivers are affected by the natural processes such as dissolution of nutrients and erosion of minerals from the overlying rocks as well as anthropogenic influences from urban, industrial and agricultural activities.

The quality of water is a very sensitive issue and it is identified in terms of its physicochemical parameters. The particular problem in the case of water quality monitoring is the complexity associated with analyzing the large number of measured variables. The data sets contain rich information about the behavior of the water resources.

Regular drinking water monitoring is essential for supplying people with a high quality and healthy water meeting all requirements of human beings. Distribution systems are being usually monitored on many sampling points at which samples are regularly taken and then analysed in laboratories in accordance with a monitoring plan.

The water quality evaluation should be based on the statistical analysis of the collected physical, chemical, and biological results.

The water quality is mostly characterized by many variables (parameters) which represent a water composition in specific localities and time. This degradation of water quality resulted in altered species composition and decreased overall health of aquatic communities.

Water quality data are not normally distributed, often co-linear or auto-correlated, containing outliers or errors etc. These data sets create a n-dimensional space from which information about the water composition has to be mined. A variety of methods are being used to display the information which is concealed in the quality variables observed in a water quality monitoring network. For this purpose, multivariate methods such as principal component analysis, factor analysis, cluster analysis and discriminant analysis are used.

Correlation or covariance matrices may be used in principal component analysis. The sums of squares and sums of products of the normalized scores constitute the correlation matrix.

Principal component analysis has been used for the data clustering and finding hidden relationships among them. This also gives simpler and more easily interpretable results for the evaluation of observed quality data. PCA also have been applied for identification factors that influence water systems for reliable management of water resources as well as for rapid solution for pollution problems

2. Basics of Principal Component Analysis

PCA as the multivariate analytical tool is used to reduce a set of original variables and to extract a small number of latent factors (principal components, PCs) for analyzing relationships among the observed variables.

The purpose of PCA is usually to determine a few linear combinations of the original variables, which can be used for summarizing the data with minimal loss of information. Principal components were computed by factors that are the orthogonal principal components (eigenvectors) of the correlation matrix of the original data. The eigenvector with the highest eigen value is the principle component of the data set.

Each principal component (PC) is a linear combination of the original variables and describes a different source of variation (information)

$$PC_i = w_1x_1 + w_2x_2 + \ldots + w_nx_n$$

Where x_i and w_i are the original variable and the component weight, respectively. The principal component weights are used as measures of the correlation between the variables and the principal components. PCA is designed to transform the original variables into new, uncorrelated variables (axes), called the principal components, which are linear combinations of the original variables.

The largest or first PC is oriented in the direction of largest variation of the original variables and passes through the centre of the data. The new axes lie along the directions of maximum variance.

The second largest PC lies in the direction of the next largest variation, passes through the centre of the data and is orthogonal to the first PC. PCA provides an objective way of finding indices of this type so that the variation in the data can be accounted for as concisely as possible.

The third largest PC is directed towards the next largest variance, goes through the data centre and is orthogonal to the first and second PCs, and so forth.

PCA starts by building the correlation matrix for the data. Diagonalization of the matrix provides its eigen values and eigen vectors. Since the variance explained by each eigen vector is proportional to its eigen value. Only those eigen vectors with eigen values greater than 1 are selected as significant independent variables (components). Sum of eigen values is equal to the total number of variables.

PCA concerned with elucidating the covariance structure of a set of variables. In particular it allows us to identify the principal directions in which the data varies. In computational terms the principal components are found by calculating the eigenvectors and eigen values of the data covariance matrix. This process is equivalent to finding the axis system in which the co-variance matrix is diagonal. The eigenvector with the largest eigen value is the direction of greatest variation, the one with the second largest eigen value is the (orthogonal) direction with the next highest variation and so on. Correlation of principal components and original variables is called loadings. The eigen vectors or components are more easily interpretable if there is a VARIMAX rotation, Which transfers the eigen vectors to make each of the representative of individual source of variation, is applied. Then each component may be identified as a source of pollution by determining its most interrelated parameters.

A detailed water quality data set needs to perform in order to get clearer image in complex data. It is a prominent technique for pattern recognition in attempts to explain the variance of a large set of inter-correlated variables and transforming into a smaller set of independent (uncorrelated) variables (principal components).

3. A Variable Reduction Procedure

Water quality assessment is the periodic measurement of multiple parameters in different monitoring stations which resulted in a complex data matrix of a large number of physico-chemical parameters that should be assessed to evaluate water quality. The use of PCA to water quality assessment has increased in recent years, mainly due to the need to obtain appreciable data reduction for analysis and decision. PCA with the aim to group the individual parameter components by the loading plots for the investigated contaminated sites.

Principle component analysis aims to uncover a more underlying set of factors that accounts for the major pattern across all the original variables.

Principal Component analysis is very useful in the analysis of data corresponding to a large number of variables. Analysis via these techniques produces easily interpretable results. It consists of observations on several variables for a number of samples or sample vectors. PCA is often applied for the removal of data noise by the reduction of their dimensionality.

PCA, based on the decomposition of a covariance/correlation matrix by the eigen value decomposition or by the singular value decomposition of real data matrices. For multivariate data, covariance is a measure of the relationship between different variables, or dimensions of the data set. The intrinsic values of water quality data are inadequate for the investigation of multivariate data table as the variables are correlated. Therefore, the indirect relationship between analytical parameters should be taken in to account for complete understanding of surface water quality.

The eigen values or singular values indicate variations among the observed variables (parameters). PCA approaches allow deriving hidden information from the data set about the possible influences of the environment on water quality.

The water quality is mostly characterized by many variables (parameters) which represent a water composition in specific localities and time. PCA is a robust statistical technique to reduce data and develop composite variables. Water quality is measured by several key indicators, but its overall composition and spatial distribution is often difficult to discern.

To assess the underlying patterns in the distribution of the measured parameters, principal component analysis (PCA) was used to extract composite variables (principal components) from the original data.

The PCA solution was rotated (using VARIMAX) to facilitate the interpretation of the principal components, and the factor scores were saved for each data record. Spatial distribution of the mean and SD of the factor scores for each principal component.

The analysis of variant species of pollutants is more advantageous than a single one, where PCA was helpful to reduce and extract the most effective groups of environmental pollutants and also to assign water quality within areas under investigation. Thus offers an effective early warning system for environmental monitoring programs.

It illustrates the most significant parameters, which describe whole data set rendering data reduction with minimum loss of original information. PCA are sensitive to outliers, missing data and poor linear correlation between variables due to inadequate assigned variables.

4. Characteristics of principal components

The first component extracted in a principal component analysis accounts for a maximal amount of total variance in the observed variables.

Under typical conditions, this means that the first component will be correlated with at least some of the observed variables. It may be correlated with many. The second component extracted will have two important characteristics. First, this component will account for a maximal amount of variance in the data set that was not accounted for by the first component. Again under typical conditions, this means that the second component will be correlated with some of the observed variables that did not display strong correlations with component 1.

The second characteristic of the second component is that it will be uncorrelated with the first component. Literally, if you were to compute the correlation between components 1 and 2, that correlation would be zero.

The remaining components that are extracted in the analysis display the same two characteristics:

Each component accounts for a maximal amount of variance in the observed variables that was not accounted for by the preceding components, and is uncorrelated with all of the preceding components. When the analysis is complete, the resulting components will display varying degrees of correlation with the observed variables, but are completely uncorrelated with one another. The extracted uncorrelated components are called principal components (PC) and are estimated from the eigenvectors of the covariance or correlation matrix of the original variables.

5. Method of data processing

The water quality parameters obtained from the laboratory analysis were used as variable inputs for principal components analysis (PCA). The data were standardized to produce a normally distribution of all variables. Since water quality parameters had different magnitudes and scales of measurements, which if not taken into account would have given more weight to certain variables due to their respective variance.

This PCA technique aims to transform the observed variables to a new set of variables. Principal components (PCs) which are uncorrelated and arranged in decreasing order of importance to simplify the problem.

In PCA, data cluster is rotated by subtracting the mean of the data and dividing by the standard deviation. So the centroid of whole data set is zero and relative location of all the points remains the same. This type of ordination reduces the dimensionality of the data set and minimizes the loss of information caused by reduction.

From the standardized covariance or correlation matrix of the data, the eigen values and corresponding eigen vectors of covariance matrix were calculated. Then a number of PCs were selected from the initial PC according to their eigen values and scree diagram. The eigen value associated with each principal component tells us how much variation in the data set it explains. They are usually expressed as a percentage of the total variation in the data set.

6. Pretreatment Data Set

Principal component analysis was performed on correlation matrix of rearranged data (all observations for each water quality data of sites), thus explains the structure of the underlying data set.

The correlation coefficient matrix measures how well the variance of each constituent can be explained based on the relationship with each others. PCA of the normalized variables (water quality data) was performed to extract significant PCs and to further reduce the contribution of variables with minor significance; these PCs were subjected to varimax rotation generating VFs.

The principal components resulted by PCA are sometimes not readily interpreted and varimax rotation need to perform to reduce the dimensionality of the data and identify most significant new variables.

Chemical data were analyzed by the principal component analysis, which quantifies relationship between the variables by computing the matrix of correlations for the entire dataset.

7. Rotation of Principal Components

In principal components analysis, the variables are rotated to obtain new variables (principal components or principal axes) and later the number of principal components is reduced by eliminating some relatively unimportant components. Sometimes the first few principal components selected are rotated to achieve a new set of components which can be more easily interpreted.

A variety of rotation techniques (varimax, equamax, quartimax) may be used for this purpose. Varimax rotation is the most widely used rotation in principal component analysis. This rotation, which includes an orthogonal rotation, is too complicated a technique to explain data.

The idea is that each variable should be heavily loaded on as few components as possible. One such technique for accomplishing this transformation is a varimax rotation. This technique tends to eliminate medium-range correlations between the components and the original variables, thus simplifying the decision as to which of the original variables to include in the components extracted

8. Identification of Important Components

Components loading (correlation coefficients), which measure the degree of closeness, between the variables and the PC, the largest loading either positive or negative, suggest the meaning of the dimensions; positive loading indicates that the contribution of the variables increases with the increasing loading in dimension; and negative loading indicates a decrease.

After computing the variances (eigen values, or latent roots) and principal components (eigenvectors) of a correlation (or covariance) matrix, the usual procedure is to look at the first few components which may be account for a large proportion of the total variance.

An eigen value gives a measure of the significance of the factor: the factors with the highest eigen values are the most significant. Eigen values of 1.0 or greater are considered significant.

9. Factor Loading

FA follows PCA. The main purpose of FA is to reduce the contribution of less significant variables to simplify even more of the data structure coming from PCA. This purpose can be achieved by rotating the axis defined by PCA, according to well-established rules, and constructing new variables, also called varifactors (VF).

PC is a linear combination of observable water quality variables, whereas VF can include unobservable, hypothetical, latent variables.

The contribution of the original variables to the extracted principal component is proportional to the linear combinations of original variables. These contributions are called "Factor Loading". When several variables have large loadings on a principal component, they may be interpreted as being strongly associated with each other and significant in that principal component.

On the other hand, a small loading of variables on a principal component indicates that the variable is not associated with the principal component.

Principal component analysis and factor analysis have been used for water quality assessment, which are under the influences of many factors during the chemical analysis.

The PCA methods was applied to the chemical concentration data in order to study the geochemistry variables capable of promoting a characterization of the hydrochemistry of the region and to identify the fundamental factors that govern the general behavior of the water sources.

The use of PCA may be a useful tool in reducing the complexity of the data and revealing spatial distribution of the water quality indicators.

10. Importance of Principal Components Analysis in revealing the variations of water Quality

Principal component analysis method (PCA) is a comparative complete method of multivariate statistical analysis, which can force the multidimensional indexes to be in one system and then study these data quantificationally. Because of those advantages, it has applied to evaluate water quality.

PCA is used to find a few comprehensive indexes, which have great influence on water quality, by studying the internal structure of correlation matrix of initial variable. It not only remains the initial main information, but also keeps the indexes from outside disturbance.

PCA makes it easier to grasp principal contradictions when dealing with the complex system of water quality. PCA doesn't need to be scored by expert, which makes data processing more objective and scientific. The principal components of water quality are controlled by lithology, gentle slope gradient, poor drainage, long residence of water, ion exchange, weathering of minerals, heavy use of fertilizers, and domestic wastes. Varieties of approaches are being used to interpret the concealed variables that determine the variance of observed water quality of various sources. Data were initially arranged according to the stations and year of monitoring. Any particular variable that has not been detected (below detection limit), the value are normally set to half and no missing data was ensure in the overall data sets.

Several data that are not normally distributed were pretreated which is a combination of centering, standardization and log-scaling method. Standardization opts was implemented to increase the influence of variables whose variance is small and vice versa. Log scaling is very common in environmental data since some of the variables might exhibit too low or high values.

Principal component analysis has been successfully applied to sort out hydro-geochemical processes from commonly collected ground water quality data. To establish the spatial and temporal variations in water quality, regular monitoring programs are required, thus PCA is being used in this study.

The PCA methods was applied to the chemical concentration data in order to study the physicochemical variables, heavy metals and some organic pollutants capable of promoting a characterization of the hydrochemistry of the region and to identify the fundamental factors that govern the general behaviour of the polluted water sources.

11. Application of Principal Component Analysis (PCA)

Principal component analysis (PCA) is the method that provides a unique solution, so that the original data can be reconstructed from the results. It has been widely applied in environmental data reduction and interpretation of multi-constituent chemical, physical biological measurements.

Principal components (PCs) actually take the cloud of data points and rotate it such a way that maximum variability is visible. In other words, it identifies the most important gradients. In recent years many studies have been done using PCA in the interpretation of water quality parameters.

Principal component often present information on the most meaningful reliable parameters, which define the whole data set affording data reduction with minimum loss of original information.

This technique aims to transform the observed variables to a new set of variables which are uncorrelated and arranged in decreasing order of importance.

The principal aim is to simplify the problem and to find new variables (principal components) which make the data easier to understand.

Principal component analysis employed to investigate the factors which caused variations in the observed quality data. This study also demonstrates the usefulness of the technique in the analysis of water quality data.

It stated that most applications of this analysis have involved a correlation matrix rather than a covariance matrix. They stated that if the parameters (variables) are in widely different units (mg/l, pH, ^{0}C etc.), then standard variants and correlation matrix should be used.

It illustrates the usefulness of principal component analysis, factor analysis and discriminant analysis for the analysis and interpretation of complex datasets and in water quality assessment, identification of pollution sources/factors, and understanding of temporal and spatial variations of water quality for effective river water quality management.

The results of principal components analysis demonstrate that it provides reliable information with respect to reality in fields of scientific research.

12. Interpretation of the Results

Correlation coefficients and the eigen values regarding the components were computed for all the variables at the first step. The proportion of the total variance explained by each principle component is additive, with each new component contributing less than the preceding one to the explained variance i.e. the components are derived in decreasing order of importance.

Subsequently, these components were rotated to eliminate medium-range loadings (correlations) to make the interpretation of the components easier.

Components which explain a relatively small proportion of the total variance of the principal components, the most significant variables in the components represented by high loadings have been taken into consideration in evaluating the components.

In addition to the significance of high loading values, there exists a difference between the components; the components with larger variances are more desirable since they give more information about the data.

13. Conclusion

The objective of PCA is to achieve parsimony and reduce dimensionality by extracting the smallest number components that account for most of the variation in the original multivariate data and to summarize the data with little loss of information.

PCA has been applied for the assessment of the seasonal and polluting effects on water quality, evaluation of monitoring stations, temporal and spatial variations on water quality and pollution source identification.

PCA was applied to assess the water quality of rivers impacted by organic pollution. The combination of both environmental variables and biological indicators in PCA was able to clearly define and explain the pollution gradients.

Therefore, the use of PCA to detect changes in community structure, ecological integrity of surface water systems and environmental impact assessment will enable to understand an integrative health condition of surface water systems due to chemical, physical and biological stressors.

Principal component analysis is a powerful tool for reducing a number of observed variables into a smaller number of artificial variables that account for most of the variance in the data set.

It is particularly useful when you need a data reduction procedure that makes no assumptions concerning an underlying causal structure that is responsible for covariation in the data. When it is possible to postulate the existence of such an underlying causal structure, it may be more appropriate to analyze the data using exploratory factor analysis.

The usual program of water quality assessment depends on the periodic measurement of multiple parameters in different monitoring stations which resulted in a complex data matrix of a large number of physico-chemical parameters. Therefore, to simplify the problem of data reduction and draw meaningful conclusion multivariate statistical techniques have been widely used.

In this regard, PCA, a powerful multivariate statistical technique, is applied to reduce the dimensionality of a data set while retaining as much as possible the variability present in data set and allows to assess associations between variables.

The analysis of variant species of pollutants is more advantageous than a single one, where PCA was helpful to reduce and extract the most effective groups of environmental pollutants and also to assign water quality within areas under investigation. Thus offers an effective early warning system for environmental monitoring programs.

14. References

- Vega, M., Pardo, R., Barrado, E. & Deban, L. 1998 Assessment of seasonal and polluting effects on the quality of river water by exploratory data analysis. Wat. Res. 32, 3581–3592.

- V. Simeonov, J.W. Einax, I. Stanimirova, , & J.Kraft, 2002. *Analytical and Bioanalytical Chemistry*, 2002,374,898.

- Sargaonkar, &V. Deshpande, .*Environmental Monitoring and Assessment*, 2003, 89, 43.

- Shashikant R. K uchekat et al., 2011. *Der chemica Sinica*, **2011**,2(4),229

- Yadav S.S and Kumar Rajesh, 2011. *Adv. Appl. Sci. Res.* 2011,2(2):197

- Reghunath, R.; Murthy, T. & Raghavan, B. (2002). The utility of multivariate statistical techniques in hydrogeochemical studies: an example from Karnataka, India. *WaterResearch*, Vol. 36, No. 10, pp. 2437-2442.

- Dr. Parinya Sanguansat, Principal Component Analysis - Engineering Applications, (Ed.), ISBN: 978-953-51-0182-6.

- Dr. Hemant Pathak, Multivariate Statistics: An approach for water quality assessment, Lambert Academic Publication, Germany, ISBN: 978-3-8454-2367-8

- Dr. Hemant Pathak, Study of seasonal variation in ground Water quality Chemical parameters of Sagar city (M.P.) by principal component analysis and evaluation, vol. 8(4), E- Journal of chemistry,2011, ISSN: 0973-4945, www.ejchem.net/PDF/V8N4/2000-2009.pdf

- Dr. Hemant Pathak, Doctoral thesis- Mathematical modeling of environmental impact assessment in reference to water soluble chemical waste (2012), (awarded), Dr. H. S. Gour central university, Sagar, M.P., India.

- Mardia, K. V., Kent, J. T. and Bibby, J. M.: 1979, *Multivariate Analysis*. Academic Press, New York, pp. 213–254.

- Sukarma T., Siddhartha C. and Priyanka T., (2011): Assessment of Water Quality of Ganga River in Kanpur by Using Principal Components Analysis. Advances in Applied Science Research, 2 (5):84-91.

- Shihab A., (1993): Application of Multivariate Method in the Interpretation of water Quality Monitoring Data of Saddam Dam Reservoir, 13.

www.ingramcontent.com/pod-product-compliance
Lightning Source LLC
Chambersburg PA
CBHW081250170526
45165CB00009B/3267